An Idea Can Go Extinct

1. Greta Thunberg *No One Is Too Small to Make a Difference*
2. Naomi Klein *Hot Money*
3. Timothy Morton *All Art is Ecological*
4. George Monbiot *This Can't Be Happening*
5. Bill McKibben *An Idea Can Go Extinct*
6. Amitav Ghosh *Uncanny and Improbable Events*
7. Tim Flannery *A Warning from the Golden Toad*
8. Terry Tempest Williams *The Clan of One-Breasted Women*
9. Michael Pollan *Food Rules*
10. Robin Wall Kimmerer *The Democracy of Species*
11. Dai Qing *The Most Dammed Country in the World*
12. Wangari Maathai *The World We Once Lived In*
13. Jared Diamond *The Last Tree on Easter Island*
14. Wendell Berry *What I Stand for Is What I Stand On*
15. Edward O. Wilson *Every Species is a Masterpiece*
16. James Lovelock *We Belong to Gaia*
17. Masanobu Fukuoka *The Dragonfly Will Be the Messiah*
18. Arne Naess *There is No Point of No Return*
19. Rachel Carson *Man's War Against Nature*
20. Aldo Leopold *Think Like a Mountain*

An Idea Can Go Extinct

BILL McKIBBEN

PENGUIN BOOKS — GREEN IDEAS

PENGUIN BOOKS

UK | USA | Canada | Ireland | Australia
India | New Zealand | South Africa

Penguin Books is part of the Penguin Random House group of companies
whose addresses can be found at global.penguinrandomhouse.com.

The End of Nature first published by Random House Inc. in 1989
This extract published in Penguin Books 2021

001

Set in 12/14.5pt Dante MT Std
Typeset by Jouve (UK), Milton Keynes
Printed and bound in Great Britain by Clays Ltd, Elcograf S.p.A.

The authorized representative in the EEA is Penguin Random House Ireland,
Morrison Chambers, 32 Nassau Street, Dublin D02 YH68

A CIP catalogue record for this book is available from the British Library

ISBN: 978–0–241–51441–2

www.greenpenguin.co.uk

Almost every day, I hike up the hill out my back door. Within a hundred yards the woods swallow me up, and there is nothing to remind me of human society – no trash, no stumps, no fence, not even a real path. Looking out from the high places, you can't see road or house; it is a world apart from man. But once in a while someone will be cutting wood farther down the valley, and the snarl of a chain saw will fill the woods. It is harder on those days to get caught up in the timeless meaning of the forest, for man is nearby. The sound of the chain saw doesn't blot out all the noises of the forest or drive the animals away, but it does drive away the feeling that you are in another, separate, timeless, wild sphere.

Now that we have changed the most basic forces around us, the noise of that chain saw will always be in the woods. We have changed the atmosphere, and that will change the weather. The temperature and rainfall are no longer to be entirely the work of some separate, uncivilizable force, but instead in part a product of our habits, our economies, our ways of life. Even in the most remote wilderness, where the strictest laws forbid the felling of a single tree, the sound of that saw will be clear, and a walk in the woods will be changed – tainted – by its whine. The world outdoors will mean much the same thing as the world indoors, the hill the same thing as the house.

An idea, a relationship, can go extinct, just like an animal or a plant. The idea in this case is 'nature,' the separate and wild province, the world apart from man to which he adapted, under whose rules he was born and died. In the past, we spoiled and polluted parts of that nature, inflicted environmental 'damage.' But that was like stabbing a man with toothpicks: though it hurt, annoyed, degraded, it did not touch vital organs, block the path of the lymph or blood. We never thought that we had wrecked nature. Deep down, we never really thought we could: it was too big and too old; its forces – the wind, the rain, the sun – were too strong, too elemental.

But, quite by accident, it turned out that the

carbon dioxide and other gases we were producing in our pursuit of a better life – in pursuit of warm houses and eternal economic growth and of agriculture so productive it would free most of us from farming – *could* alter the power of the sun, could increase its heat. And that increase *could* change the patterns of moisture and dryness, breed storms in new places, breed deserts. Those things may or may not have yet begun to happen, but it is too late to altogether prevent them from happening. We have produced the carbon dioxide – we are ending nature.

We have not ended rainfall or sunlight; in fact, rainfall and sunlight may become more important forces in our lives. It is too early to tell exactly how much harder the wind will blow, how much hotter the sun will shine. That is for the future. But the *meaning* of the wind, the sun, the rain – of nature – has already changed. Yes, the wind still blows – but no longer from some other sphere, some inhuman place.

In the summer, my wife and I bike down to the lake nearly every afternoon for a swim. It is a dogleg Adirondack lake, with three beaver lodges, a blue heron, some otters, a family of mergansers, the occasional loon. A few summer houses cluster at one end, but mostly it is surrounded by wild state land. During the week we swim across and back, a trip of maybe forty minutes – plenty of time to forget everything but the feel of the water around your body and the rippling, muscular joy of a hard kick and the pull of your arms.

But on the weekends, more and more often, someone will bring a boat out for waterskiing, and make pass after pass up and down the lake. And then the whole experience changes, changes entirely. Instead of being able to forget everything but yourself, and even yourself except for the muscles and the skin, you must be alert, looking up every dozen strokes to see where the boat is, thinking about what you will do if it comes near. It is not so much the danger – few swimmers, I imagine, ever die by Evinrude. It's not even so much the blue smoke that hangs low over the water. It's that the motorboat gets in your mind. You're forced to think, not feel – to think of human society and of people. The lake is utterly different on these days, just as the planet is utterly different now.

The argument that nature is ended is complex; profound objections to it are possible, and I will try to answer them. But to understand what's ending requires some attention to the past. Not the ancient past, not the big bang or the primal stew. The European exploration of this continent is far enough back, for it is man's *idea* of nature that is important here, and it was in response to this wild New World that much of our modern notion of nature developed. North America, of course, was not entirely unaltered by man when the colonists arrived, but its previous occupants had treated it fairly well. In many places, it was wilderness.

And most of it was wilderness still on the eve of the Revolution, when William Bartram, one of America's first professional naturalists, set out from his native Philadelphia to tour the South. His report on that trip through 'North and South Carolina, Georgia, East and West Florida, the Cherokee Country, the Extensive Territories of the Muscogulges, or Creek Confederacy, and the Country of the Choctaws' is a classic; it gives the sharpest early picture of this fresh continent. Though some of the land he traveled had been settled (he spent a number of his nights with gentlemen farmers on their plantations), the settlement was sparse, and the fields of indigo and rice gave way quickly to the wilderness. And not the dark and forbidding wilderness of European fairy tales but a blooming, humming, fertile paradise. Every page of Bartram's long journal shouts of the fecundity, the profligacy, of this fresh land. 'I continued several miles [reaching] verdant swelling knolls, profusely productive of flowers and fragrant strawberries, their rich juice dyeing my horse's feet and ankles.' When he stops for dinner, he catches a trout, picks a wild orange, and stews the first in the juices of the second over his fire.

Whatever direction he struck off in, Bartram found vigorous beauty. He could not even stumble in this New World without discovering something: near the Broad River, while ascending a 'steep, rocky hill,' he slips and reaching for a shrub to steady himself he tears up several plants of a new species of *Caryophyllata (geum odoratissimum)*. Fittingly, their roots 'filled the air with animating scents of cloves and spicy perfumes.' His diary brims over with the grand Latin binomials of a thousand plants and animals – *Kalmia latifolia*, 'snowy mantled' *Philadelphus inodorus, Pinus sylvestris, Populus tremula, Dionea muscipula* ('admirable are the properties' of these 'sportive vegetables'!), *Rheum rhubarbarum, Magnolia grandiflora* – and also with the warm common names: the bank martin, the water wagtail, the mountain cock, the chattering plover, the bumblebee. But the roll call of his adjectives is even more indicative of his mood. On one page, in the account of a single afternoon, he musters fruitful, fragrant, sylvan (twice), moderately warm, exceedingly pleasant, charming, fine, joyful, most beautiful, pale gold, golden, russet, silver (twice), blue green, velvet black, orange, prodigious, gilded, delicious, harmonious, soothing, tuneful, sprightly, elevated, cheerful (twice), high and airy, brisk and cool, clear, moonlit, sweet, and healthy. Even where he can't see, he imagines marvels: the fish disappearing into subterranean streams, 'where, probably, they are separated from each other by innumerable paths, or secret rocky avenues, and after encountering various obstacles, and beholding new and unthought-of scenes of pleasure and disgust, after many days' absence from the surface of the world emerge again from the dreary vaults, and appear exulting in gladness and sporting in the transparent waters of some far distant lake.' But he is no Disney – this is no *Fantasia*. He is a scientist recording his observations, and words like 'cheerful' and 'sweet' seem to have been technical descriptions of the untouched world in which he wandered.

This sort of joy in the natural was not a literary

convention, a given; as Paul Brooks points out in *Speaking for Nature*, much of literature had regarded wilderness as ugly and crude until the Romantic movement of the late eighteenth century. Andrew Marvell, for instance, referred to mountains as 'ill-designed excrescences.' This silliness changed into a new silliness with the Romantics; Chateaubriand's immensely popular *Atala*, for instance, describes the American wilderness as filled with bears 'drunk with grapes, and reeling on the branches of the elm trees.' But the rapturous fever took on a healthier aspect in this country. If most of the pioneers, to be sure, saw a buffalo as something to hunt, a forest as something to cut down, a flock of passenger pigeons as a call to heavy artillery (farmers would bring their hogs to feed on the carcasses of pigeons raining down in the slaughter), there were always a good many (even, or especially, among the hunters and loggers) who recognized and described the beauty and order of this early time.

Of a thousand examples, my favorite single description comes from George Catlin, who traveled across the frontier to paint the portraits of American Indians. In his journal he describes a night he spent while riding north from Fort

Gibson to the Missouri River in order to escape an epidemic. His camp was 'in one of the most lovely little valleys I ever saw, and even far more beautiful than could be imagined by mortal man,' he writes, 'an enchanting little lawn of five or six acres, on the banks of a cool and rippling stream, that was alive with fish; and every now and then, a fine brood of ducks, just old enough for delicious food and too unsophisticated to avoid an easy and simple death. This little lawn was surrounded by bunches and copses of the most picturesque foliage, consisting of lofty bois d'arcs and elms, spreading their huge branches as if offering protection to the rounded groups of cherry and plum trees that supported festoons of grapevines with their purple clusters that hung in the most tempting manner over the green carpet that was everywhere decked out with wild flowers of all tints and various sizes, from the modest wild sunflowers, with their thousand tall and droopy heads, to the lilies that stood, and the violets that crept, beneath them . . . The wild deer were repeatedly rising from their quiet lairs, and bounding out and over the grace-full swells of the prairies which hemmed it in.' If this passage had a little number at the start of each sentence, it could be Genesis; it sticks in my mind as a baseline, a reminder of where we began.

Such visions of the world as it existed outside human history became scarcer with each year that passed, of course. By the 1930s, when Bob Marshall, the founder of the Wilderness Society, set off to explore Alaska's Brooks Range, all the lower forty-eight states had been visited, mapped, and named. 'Often, as when visiting Yosemite or Glacier Park or the Grand Canyon or Avalanche Lake or some other natural scene of surpassing beauty, I had wished selfishly enough that I might have had the joy of being the first person to discover it,' he wrote. 'I had been thrilled reading Captain Lewis's glowing account of the great falls of the Missouri. I yearned for adventures comparable to those of Lewis and Clark.' And he found them, on the upper reaches of the Koyukuk River, where no one, not even an Alaskan Eskimo, seems ever to have been before. Each day brought eight, ten, a dozen ridges and streams and peaks under his eye and hence into human history. One morning he came around a corner to discover that 'the Clear River emerged from none of three gorges we had imagined, but from a hidden valley which turned almost at right angles to the west. I cannot convey in words my feeling in finding this broad valley lying there, just as fresh and untrammeled as at the dawn of geological eras hundreds of millions of years ago. Nor is there any adequate way of describing the scenery . . . I could make mention of thousand-foot sheer precipies; I could liken the valley to a Yosemite without waterfalls, but with rock domes beside which the world-renowned Half Dome would be trivial – yet with all that I would not have conveyed the sense of the continuous, exulting feeling of immensity . . . Best of all it was fresh – gloriously fresh. At every step there was the exhilarating feeling of breaking new ground. There were no musty signs of human occupation. This, beyond a doubt, was an unbeaten path.'

Marshall was very near the last to see surroundings unpolluted even by the knowledge that someone had been there before. Though his explorations were made not long before World War II, they were an anomaly, a last gasp of the voyages of discovery that marked an earlier epoch. It is hard for us to credit that only a hundred and thirty years ago the valley of the Colorado – the Grand Canyon – was a blank spot on maps of the Southwest, or that sixty years before that the Rockies were a rumor among white men. That when Thoreau climbed Maine's Mt Katahdin in 1846 he could list the names of the five Europeans who had preceded him up the peak. 'I am reminded by my journey how exceedingly new this country is,' Thoreau wrote. 'Those Maine woods differ essentially from ours [in Concord, Massachusetts]. There you are never reminded that the wilderness which you are threading is, after all, some villager's familiar wood-lot, some widow's thirds, from which her ancestors have sledded fuel for generations, minutely described in some old deed.' Here in the Adirondacks, our highest peak, Mt Marcy, was not climbed by a white man until 1837, a generation after the return of Lewis and Clark.

We are rarely reminded anymore of the continent's newness. That era of discovery is as firmly closed to us as the age of knights and dragons. Katahdin, though preserved as a park, is so popular that the authorities must strictly limit the number of campers – some days hundreds are at the summit simultaneously. The trail up Mt Marcy on a holiday weekend is like the Macy's escalators with a heavy balsam scent. I once interviewed a man who was *rowing* to Antarctica from Tierra del Fuego because, he explained, 'you can't be the first to explore the blank spots on the map or to climb the mountains anymore. It has a lot more to do with style now.' (He had previously skied *around* Mt Everest.) Not even the moon to conquer!

Over time, though, we've reconciled ourselves to the idea that we'll not be the first up any hill, and, indeed, we've come to appreciate the history of a spot as a source of added pleasure and interest. On the prairies we search for the rutted tracks left by the wagon trains; at Walden Pond, where Thoreau sought to escape man, we dutifully trek around the shore to see the site of his cabin. In something of the same fashion, we have come to accept, and enjoy, the intrusion of scientific explanation – to know that we can marvel with undiminished awe at the south wall of the Grand Canyon even while understanding the geologic forces that carved it. The Grand Canyon is so . . . grand that we can cope with not being the first people to see it. The wonder of nature does not depend on its freshness.

But still we feel the need for pristine places, places substantially *unaltered* by man. Even if we do not visit them, they matter to us. We need to know that though we are surrounded by buildings there are vast places where the world goes on as it always has.

The Arctic National Wildlife Refuge, on Alaska's northern shore, is reached by just a few hundred people a year, but it has a vivid life in the minds of many more, who are upset that oil companies want to drill there. And upset not only because it may or may not harm the caribou but because here is a vast space free of roads and buildings and antennas, a blank spot if not on the map then on the surface. It sickens us to hear that 'improper waste disposal practices' at the American Antarctic research station in McMurdo Sound have likely spread toxic waste on that remote continent, or that an Exxon tanker has foundered off the port of Valdez, tarring the beaches with petroleum.

One proof of the deep-rooted desire for pristine places is the decision that Americans and others have made to legislate 'wilderness' – to set aside vast tracts of land where, in the words of the federal statute, 'the earth and its community of life are untrammeled by man, where man himself is a visitor who does not remain.' Pristine nature, we recognize, has been overwhelmed in many places, even in many of our national parks. But in these few spots it makes a stand. If we can't have places where no man has ever been, we can at least have spots where no man is at the moment.

Segregating such wilderness areas has not been easy. The quiet of the land behind my house, fifty thousand acres of state wilderness, is daily broken by Air Force jets practicing flying beneath radar; they come in pairs, twisting and screeching above the hills, so that for a moment, and a few moments after that, it is no wilderness at all. And often, of course, man invades more insidiously: the synthetic compounds of man's pesticides, for instance, worm their way slowly but inevitably into the fabric of life.

But, even under such stress, it is still wilderness, still pristine in our minds. Most of the day, the sky above my mountain is simply sky, not 'airspace.' Standing in the middle of a grimy English mill town, George Orwell records this 'encouraging' thought: 'In spite of hard trying, man has not yet succeeded in doing his dirt everywhere. The earth is so vast and still so empty that even in the filthy heart of civilization you find fields where the grass is green instead of grey; perhaps if you looked for them you might even find streams with live fish in them instead of salmon tins.' When Rachel Carson wrote *Silent Spring*, she was able to find some parts of the Arctic still untouched – no DDT in the fish, the beaver, the beluga, the caribou, the moose, the polar bear, the walrus. The cranberries, the strawberries, and the wild rhubarb all tested clean, though two snowy owls, probably as a result of their migrations, carried small amounts of the pesticide, as did the livers of two Eskimos who had been away to the hospital in Anchorage.

In other words, as pervasive a problem as DDT was, and is, one could, and can, always imagine that *somewhere* a place existed free of its taint. (And largely as a result of Carson's book there are more and more such places.) As pervasive and growing as the problem of acid rain surely is, at the moment places still exist with a rainfall of an acceptable, 'normal' pH. And if we wished to stop acid rain we could; experimenters have placed tents over groves of trees to demonstrate that if the acid bath ceases, a forest will return to normal. Even the radiation from an event as nearly universal as the explosion at the Chernobyl plant has begun to fade, and Scandinavians can once more eat their vegetables.

We can, in other words, still plausibly imagine wild nature – or, at least, the possibility of wild nature in the future – in all sorts of places.

This idea of nature is hardy. Our ability to shut the destroyed areas from our minds, to see beauty around man's degradation, is considerable. A few years ago I spent some days driving around Arizona in a van with a man named Lyn Jacobs, one of a small number of environmentalists fighting a difficult battle to restrict the grazing of cattle on public lands in the West. The cows, which range over 70 percent of the federal land in the American West under a leasing program that does not pay for itself and each year requires tax subsidies, produce about 3 percent of America's beef. And by their constant grazing, the cattle convert the rangelands into barren pastures. Where there are streams they cave in the banks; where there are wildfowl they trample their nests. In their wake they leave stands of cheat-grass and thistle in place of the natural long-stemmed grasses. But the West has been a pasture so long that practically no one notices. People just assume that grass there can't grow more than a foot high. One morning, Jacobs and I drove along a ranch road that ran just parallel to the Grand Canyon about fifteen miles from the south rim. It was a glorious day, the sky a polarized blue, and though you couldn't see the canyon you knew with heart-stopping precision where it was, for the clouds dropped over its edge, their bottoms obscured like icebergs. 'That's the problem,' Jacobs said, stopping the van. 'When you look at Western panoramas, you don't look down – your eye is trained to think this desert is normal.

You tend to look at the mountains and the blue sky above them, and the clouds.'

The idea of wildness, in other words, can survive most of the 'normal' destruction of nature. Wildness can survive in our minds once the land has been discovered and mapped and even chewed up. It can survive all sorts of pollution, even the ceaseless munching of a million cows. If the ground is dusty and trodden, we look at the sky; if the sky is smoggy, we travel someplace where it's clear; if we can't travel to someplace where it's clear, we *imagine* ourselves in Alaska or Australia or some place where it is, and that works nearly as well. Nature, while often fragile in reality, is durable in our imaginations. Wildness, the idea of wildness, has outlasted the exploration of the entire globe. It has endured the pesticides and the pollution. When the nature around us is degraded, we picture it fresh and untainted elsewhere. When elsewhere, too, it rains acid or DDT, we can still imagine that someday soon it will be better, that we will stop polluting and despoiling and instead 'restore' nature. (And, indeed, people have begun to do just this sort of work: here in the Adirondacks, helicopters drop huge quantities of lime into lakes in order to reduce their acidity.) In our minds, nature suffers from a terrible case of acne, or even skin cancer – but our faith in its essential strength remains, for the damage always seems local.

But now the basis of that faith is lost. The idea of nature will not survive the new global pollution – the carbon dioxide and the CFCs and the like. This new rupture with nature is different not only in scope but also in kind from salmon tins in an English stream. We have changed the atmosphere, and thus we are changing the weather. By changing the weather, we make every spot on earth man-made and artificial. We have deprived nature of its independence, and that is fatal to its meaning. Nature's independence *is* its meaning; without it there is nothing but us.

If you travel by plane and dog team and snowshoe to the farthest corner of the Arctic and it is a mild summer day, you will not know whether the temperature is what it is 'supposed' to be, or whether, thanks to the extra carbon dioxide, you are standing in the equivalent of a heated room. If it is twenty below and the wind is howling – perhaps absent man it would be forty below. Since most of us get to the North Pole only in our minds, the real situation is more like this: if in July there's a heat wave in

London, it won't be a natural phenomenon. It will be a man-made phenomenon – an amplification of what nature intended or a total invention. Or, at the very least, it *might* be a man-made phenomenon, which amounts to the same thing. The storm that might have snapped the hot spell may never form, or may veer off in some other direction, not by the laws of nature but by the laws of nature as they have been rewritten, blindly, crudely, but effectively, by man. If the sun is beating down on you, you will not have the comfort of saying, 'Well, that's nature.' Or if the sun feels sweet on the back of your neck, that's fine, but it isn't nature. A child born now will never know a natural summer, a natural autumn, winter, or spring. Summer is going extinct, replaced by something else that will be called 'summer.' This new summer will retain some of its relative characteristics – it will be hotter than the rest of the year, for instance, and the time of year when crops grow – but it will not be summer, just as even the best prosthesis is not a leg.

And, of course, climate determines an enormous amount of the rest of nature – where the forests stop and the prairies or the tundra begins, where the rain falls and where the arid deserts squat, where the wind blows strong and steady, where the glaciers form, how fast the lakes evaporate, where the seas rise. As John Hoffman, of the Environmental Protection Agency, noted in the *Journal of Forestry*, 'trees planted today will be entering their period of greatest growth when the climate has already changed.' A child born today might swim in a stream free of toxic waste, but he won't ever see a natural stream. If the waves crash up against the beach, eroding dunes and destroying homes, it is not the awesome power of Mother Nature. It is the awesome power of Mother Nature as altered by the awesome power of man, who has overpowered in a century the processes that have been slowly evolving and changing of their own accord since the earth was born.

Those 'record highs' and 'record lows' that the weathermen are always talking about – they're meaningless now. It's like comparing pole vaults between athletes using bamboo and those using fiberglass poles, or dash times between athletes who've been chewing steroids and those who've stuck to Wheaties. They imply a connection between the past and the present which doesn't exist. The comparison is like hanging Rembrandts next to Warhols; we live in a postnatural world. Thoreau once said he could walk for half an hour and come to 'some portion of the earth's surface where man does not stand from one year's end to another, and there, consequently, politics are not, for they are but the cigar-smoke of a man.' Now you could walk half a year and not reach such a spot. Politics – our particular way of life, our ideas about how we should live – now blows its smoke over every inch of the globe.

About a half mile from my house, right at the head of the lake, the town has installed a street-light. It is the only one for miles, and it is undeniably useful – without it, a car or two each summer would undoubtedly miss the turn and end up in the drink. Still, it intrudes on the dark. Most of the year, once the summer people have left, there is not another light to be seen. On a starry night the Milky Way stands out like a marquee; on a cloudy night you can walk in utter pitch-black, unable to see even the dog trotting at your side. But then, around the corner, there is the streetlamp, and soon you are in its sodium-vapor circle, a circle robbed of mystery by its illumination. It's true that the bugs love the lamp; on a June night there is more wildlife buzzing around it than in any square acre of virgin forest. But it breaks up the feeling of the night. And now it is as if we had put a huge lamp in the sky, and cast that same prosaic sterile light at all times on all places.

While I was stacking wood one morning last fall I noticed a lot of ash floating through the air. 'Did you make a fire?' I asked my wife through the window. 'No,' she said. I wandered off down the road to see if it was coming from the nearest occupied house – but that's quite a way off. I finally stopped long enough to trap a piece of the ash in my fist so I could look at it. It turned out to be a bug I had never seen before – a blackflylike creature with a gray, woolly clump of something on its back that certainly looked like ash. Not man! Nature!

If only that were the case with most of the changes around us – if only all the analogies were just analogies. If only they were all figments, and the world were the same old place it had always been. But the world, the whole world, is touched by our work, even when that work is invisible.

In a famous essay, 'Sootfall and Fallout,' which was written at the height of the atmospheric atomic testing in the early 1960s, E. B. White says that the joy he always took in his newly dug garden patch 'has been spoiled by the maggots that work in the mind. Tomorrow we will have rain, and the rain falling on the garden will carry its cargo of debris from old explosions in distant places. Whether the amount of this freight is great or small, whether it is measurable by the farmer or can only be guessed at, one thing is certain: the character of rain has changed, the joy of watching it soak the waiting earth has been diminished, and the whole meaning and worth of gardens has been called into question.' Happily, we have ceased atmospheric atomic testing. Unhappily, White's words still hold true; only, now the culprits – carbon dioxide, methane, nitrous oxide, chlorofluorocarbons – are the result not of some high and distant drama, a few grand explosions, but of a billion explosions of a hundred million pistons every second, near and far and insidiously common.

We will have a hard time believing this new state of affairs. Even the most farseeing naturalists of an earlier day couldn't comprehend that the atmosphere, the climate, could be dramatically altered. Thoreau, complaining about the logging that eventually destroyed virtually every stand of virgin timber between the Atlantic and the Mississippi, said that soon the East 'would be so bald that every man would have to grow whiskers to hide its nakedness, but, thank God, the sky was safe.' And John

Muir, the Scottish-born explorer of Yosemite, wrote one day in his diary about following a herd of grazing sheep through the valley: 'Thousands of feet trampling leaves and flowers, but in this mighty wilderness they seem but a feeble band, and a thousand gardens should escape their blighting touch. They cannot hurt the trees, though some of the seedlings suffer, and should the woolly locusts be greatly multiplied, as on account of dollar value they are likely to be, then the forests too, in time, may be destroyed. Only the sky will then be safe.' George Perkins Marsh, the first modern environmentalist, knew a century ago that cutting down forests was a horrible idea, yet he said, 'The revolutions of the seasons, with their alterations of temperatures, and of length of day and night, the climate of different zones, and the general conditions and movements of the atmosphere and seas, depend upon causes for the most part cosmical, and, of course, beyond our control.'

And even as it dawns on us what we have done, there will be plenty of opportunity to forget, at least for a while, that anything has changed. For it isn't natural *beauty* that is ended; in fact, in the same way that the smog breeds spectacular sunsets, there may appear new, unimagined beauties. What will change is the meaning that beauty carries, for when we look at a sunset, we see, or think we see, many things beyond a particular arrangement of orange and purple and rose.

It is also true that this is not the first huge rupture in the globe's history. Perhaps thirty times since the earth formed, planetesimals up to ten miles in diameter and traveling at sixty times the speed of sound have crashed into the earth, releasing, according to the British scientist James Lovelock, perhaps a thousand times as much energy as would be liberated by the explosion of all present stocks of nuclear weapons. Such events, some scientists say, may have destroyed 90 percent of all living organisms. On an even larger scale, the sun has steadily increased its brightness; it has grown nearly 30 percent more luminous since life on earth began, forcing that life to keep forever scrambling to stay ahead – a race it will eventually lose, though perhaps not for some billions of years. Or consider an example more closely resembling the sharp divide we have now crossed. About two billion years ago, the microbiologist Lynn Margulis writes, the spread of certain sorts of bacteria caused, in short order, an increase in atmospheric oxygen from one part in a million to one part in just five – from .0001 percent to 21 percent. Compared to that, the increase in carbon dioxide from 280 to 560 parts per million is as the hill behind my house to Annapurna. 'This was by far the greatest pollution crisis the earth has ever endured,' Margulis writes. Oxygen poisoned most microbial life, which 'had no defense against this cataclysm except the standard way of DNA replication and duplication, gene transfer and mutation.' And, indeed, these produced the successful oxygen-synthesizing life forms that now dominate the earth.

But each of these examples is different from what we now experience, for they were 'natural,' as opposed to man-made. A pint-sized planet cracks into the earth; the ice advances; the sun, by the immutable laws of stars, burns brighter till its inevitable explosion; genetic mutation sets certain bacteria to spewing out oxygen and soon they dominate the planet, a 'strictly natural' pollution.

One can, of course, argue that the current crisis, too, is 'natural,' because man is part of nature. This echoes the views of the earliest Greek philosophers, who saw no difference between matter and consciousness – nature included everything. James Lovelock wrote some years ago that 'our species with its technology is simply an inevitable part of the natural scene,' nothing more than mechanically advanced beavers. In this view, to say that we 'ended' nature, or even damaged nature, makes no sense, since we *are* nature, and nothing we can do is 'unnatural.' This view can be, and is, carried to even greater lengths; Lynn Margulis, for instance, ponders the question of whether robots can be said to be living creatures, since any 'invention of human beings is ultimately based on a variety of processes including that of DNA replication, no matter the separation in space or time of that replication from the invention.'

But one can argue this forever and still not really feel it. It is a debater's point, a semantic argument. When I say that we have ended nature, I don't mean, obviously, that natural processes have ceased – there is still sunshine and still wind, still growth, still decay. Photosynthesis continues, as does respiration. *But we have ended the thing that has, at least in modern times, defined nature for us – its separation from human society.*

That separation is quite real. It is fine to argue, as certain poets and biologists have, that we must learn to fit in with nature, to recognize that we are but one species among many, and so on. But none of us, on the inside, quite believe it. The Sophists contrasted the 'natural' with the 'conventional' – what exists originally with what it becomes as the result of human intervention. And their distinction, filtered through Plato and Christianity and a dozen other screens, survives, because it agrees with our instinctive sense of the world. I sit writing here in my office. On the wall facing me there is a shelf of reference books – dictionaries, The *Guinness Book of World Records*, a set of encyclopedias – and a typewriter and a computer. There's another shelf of books, all dealing with American history, on my left, and, on my right, pictures of my family, a stack of mail-order catalogs for Christmas shopping, and a radio broadcasting a Cleveland performance of Ravel's Piano Concerto in D for the left hand. Visible through the window is a steep mountain with nearly a mile of bare ridge and a pond almost at the peak.

The mountain and the office are separate parts of my life; I do not really think of them as connected. At night it's dark out there; save for the streetlamp by the lake there's not a light for twenty miles to the west and thirty to the south. But in here the light shines. Its beams stretch a few yards into the night and then falter, turn to shadow, then black. In the winter it's cold out there, but in here the fire warms us until near dawn, and when it dwindles the oil burner kicks in.

What happens in here I control; what happens out there has always been the work of some independent force. That is not to say that the outside world isn't vitally important; I moved here so I could get to the mountains easily, and I think nature means a good deal even to the most inured city dweller. But it is enough for now to say that in our modern minds nature and human society are separate things. It is this separate nature I am talking about when I use the word – 'nature,' if you like.

One could also argue that we destroyed this independent nature long ago, that there's no present need for particular distress. That the day man made his first tool he irrevocably altered nature, or the day he planted his first crop. Walter Truett Anderson, in his book *To Govern Evolution*, makes the case that everything people do – including our attempts to set aside wilderness or protect endangered species – is 'one way or another human intervention.' California, his home, was permanently changed by the 1870s, he contends, when early agribusiness followed gold miners and shepherds. Technically, of course, he is correct. Any action

alters its environment – even a bird building a nest –
and it is true that we cannot, as he puts it, 'return to
a natural order untouched by human society.' But
Anderson's argument, and others like it that have
often been employed as a rationale for further altering the environment, are too broad. Independent
nature was not dead in California in 1870; in 1870,
John Muir was just beginning his sojourn in Yosemite that would yield some of the greatest hymns to
and insights into that world beyond man. As long as
some places remained free and wild, the idea of the
free and wild could live.

The invention of nuclear weapons may actually
have marked the beginning of the end of nature: we
possessed, finally, the capacity to overmaster nature,
to leave an indelible imprint everywhere all at once.
'The nuclear peril is usually seen in isolation from
the threats to other forms of life and their ecosys-
tems, but in fact it should be seen at the very center
of the ecological crisis, as the cloud-covered Everest
of which the more immediate, visible kinds of
harm to the environment are the mere foothills,'
wrote Jonathan Schell in *The Fate of the Earth*. And
he was correct, for at the time he was writing (less
than two decades ago!) it was hard to conceive of

any threats of the same magnitude. Global warming was one obscure theory among many. Nuclear weapons were unique (and they remain so, if only for the speed with which they work). But the nuclear dilemma is at least open to human reason – we can decide not to drop the weapons, and indeed to reduce and perhaps eliminate them. And the horrible power of these weapons, which has been amply demonstrated in Japan and on Bikini and under Nevada and many times in our imaginations, has led us fitfully in that hopeful direction.

By contrast, the various processes that lead to the end of nature have been essentially beyond human thought. Only a few people knew that carbon dioxide would warm up the world, for instance, and they were for a long time unsuccessful in their efforts to alert the rest of us. Now it is too late – though not too late to ameliorate – some of the changes and so perhaps to avoid the most gruesome of their consequences. But the scientists agree that we have already pumped enough gas into the air so that a significant rise in temperature and a subsequent shift in weather are inevitable.

Just how inevitable we can see from the remedies that some scientists have proposed to save us – not the remedies, like cutting fossil fuel use and saving the rain forests, that will keep things from being any worse than they need to be, but the solutions that might bring things back to 'normal.' The most natural method anyone has suggested involves growing enormous numbers of trees to take the carbon dioxide out of the air. Take, for argument's sake, a new coal-fired electric generating station that produces a thousand megawatts and operates at 38 percent thermal efficiency and 70 percent availability. To counteract just the carbon dioxide generated by that plant, the surrounding area to a radius of 24.7 kilometers would need to be covered with American sycamore trees (a fast-growing species) planted at four-foot intervals and 'harvested' every four years. It might be possible to achieve that sort of growth rate – a government forestries expert told the Senate that with genetic screening, spacing, thinning, pruning, weed control, fire and pest control, fertilization, and irrigation, net annual growth could be 'very much higher than at present.' Even if it worked, though, would this tree plantation be nature? A walk through an endless glade of evenly spaced sycamores with the weed-control chopper hovering overhead, and the irrigation pipes gurgling quietly below, represents a fundamental break with my idea of the wild world.

Other proposals get even odder. One 'futuristic idea' described in the *New York Times* springs from the brain of Dr Thomas Stix at Princeton: he proposes the possibility of using a laser to 'scrub' chlorofluorocarbons from the earth's atmosphere before they have a chance to reach the ozone layer. Dr Stix calculates that an array of infrared lasers spaced around the world could 'blast apart' a million tons of chlorofluorocarbons a year – a procedure he refers to as 'atmospheric processing.' Down at the University of Alabama, Leon Y. Sadler, a chemical engineer, has suggested employing dozens of airplanes to carry ozone into the stratosphere (others have suggested firing a continuous barrage of 'bullets' of frozen ozone, which would melt in the stratosphere). To deal with the warming problem, Columbia geochemist Wallace Broecker has considered a 'fleet of several hundred jumbo jets' to ferry 35 million tons of sulfur dioxide into the stratosphere annually to reflect sunlight away from the earth. Other scientists recommend launching 'giant orbiting satellites made of thin films' that could cast shadows on the earth, counteracting the greenhouse effect with a sort of venetian-blind effect. Certain practical problems may hamper these various solutions; Dr Broecker, for instance, admits that injecting large quantities of sulfur dioxide into the atmosphere would increase acid rain 'and give the blue sky a whitish cast.' Still they just might work. And perhaps, as Dr Broecker contends, 'a rational society needs some sort of insurance policy on how to maintain a habitable planet.' But even if they do work – even if the planet remains habitable – it will not be the same. The whitish afternoon sky blessed by the geometric edge of the satellite cloud will fade into a dusk crisscrossed by lasers. There is no way to reassemble nature – certainly not by following the suggestion of one researcher that, in order to increase the earth's reflectivity and thus cool its temperature, we should cover most of the oceans with a floating layer of white Styrofoam chips.

There are some people, perhaps many, to whom this rupture will mean little. A few years ago a group of executives went rafting down a river in British Columbia; after an accident killed five of them, one of the survivors told reporters that the party had regarded the river as 'a sort of ersatz rollercoaster.' Nature has become a hobby with us. One person enjoys the outdoors, another likes cooking, a third favors breaking into military computers over the internet. We have become in rapid order a people whose conscious need for nature is superficial. The seasons don't matter to most of us anymore except as spectacles. In my country and in many places around this part of the nation, the fair that once marked the harvest now takes place in late August, while tourist dollars are still in heavy circulation. Why celebrate the harvest when you harvest every week with a shopping cart? I am a child of the suburbs, and even though I live on the edge of the wild I have only a tenuous understanding of the natural world. I can drive past hundreds of miles of fields without ever being able to figure out what's growing in them, unless it's corn. And even farmers have a lessened feel for the world around them. The essayist Wendell Berry quotes from an advertisement for a new tractor: 'Outside – dust, noise, heat, storm, fumes. Inside – all is quiet, comfortable, safe . . . Driver dials "inside weather" to his liking . . . He pushbuttons radio or stereo-tape entertainment.'

Even this is several steps above the philosophy expressed by a mausoleum director in a full-page newspaper ad that seems to run once a week in my newspaper: 'Above-Ground. The Clean Burial. Not Underground with Earth's Disturbing Elements.' Four of his 'clean, dry, civilized' vaults are already sold out, and a fifth is under construction. While we are still alive, we do sometimes watch a nature program, an account of squid or wildebeest, usually sponsored by Mutual of Omaha. Mostly, however, we watch *L.A. Law*.

Still, the passing of nature as we have known it, like the passing of any large idea, will have its recognizable effects, both immediately and over time. In 1893, when Frederick Jackson Turner announced to the American Historical Association that the frontier was closed, no one was aware that the frontier had been the defining force in American life. But in its absence it was understood. One reason we pay so little close attention to the separate natural world around us is that it has always been there and we presumed it always would. As it disappears, its primal importance will be clearer – in the same way that some people think they have put their parents out of their lives and learn differently only when the day comes to bury them.

How will we feel the end of nature? In many ways, I suspect. If nature means Bartram's great joy at fresh and untrammeled beauty, its loss means sadness at man's footprints everywhere. But, as with the death of a person, there is more than simply loss, a hole opening up. There are also new relation-ships that develop, and strains and twists in old relationships. And since this loss is peculiar in not having been inevitable, it provokes profound questions that don't arise when a person dies.

The first of these questions, I think, has to do with God. It may seem odd to take a physical event and go straight to the metaphysical for its meaning. But, as we have seen, nature is as much an idea as a fact. And in some way that idea is connected with God. I hesitate to go further than that, for I am no theologian; I am not even certain what I mean by God. (Perhaps some theologians join me in this difficulty.)

It is not a novel observation that religion has been in decline in the modern era. Despite the recent rise of fundamentalism, the crisis of belief continues. Many people, including me, have overcome it to a greater or a lesser degree by locating God in nature. Most of the glimpses of immortality, design, and benevolence that I see come from the natural world – from the seasons, from the beauty, from the intermeshed fabric of decay and life, and so on. Other signs exist as well, such as instances of great and selfless love between people, but these, perhaps, are less reliable. They hint at epiphany, not at the eternity that nature proclaimed. If this seems a banal notion, that is exactly my point. The earliest gods we know about were animals – tigers, birds, fish. Their forms and faces peer out from ancient ruins, and from the totems and wall paintings of our first religions.

And though, as time went on, we began to give our gods human features, much feeling still adheres to the forests and fields and birds and lions – else why should we moan about the 'desecration' of our environment? I am a reasonably orthodox Methodist, and I go to church on Sunday because fellowship matters, because I find meaning in the history of the Israelites and in the Gospels, and because I love to sing hymns. But it is not in 'God's house' that I feel his presence most – it is in his outdoors, on some sun-warmed slope of pine needles or by the surf. It is there that the numbing categories men have devised to contain this mystery – sin and redemption and incarnation and so on – fall away, leaving the overwhelming sense of the goodness and the sweetness at work in the world.

Perhaps this emotion has dimmed in an urban age, and most people now perceive God through the Christian Broadcasting Network. There is no question, though, that this is one thing nature *has*

meant, and meant not just to the ancients but to the
great American naturalists who first helped us see
the outdoor world as more than a source of raw
materials or the home of dangerous animals. 'We
now use the word Nature very much as our fathers
used the word God,' John Burroughs wrote at the
turn of the century, 'and, I suppose, back of it all we
mean the power that is everywhere present and
active, and in whose lap the visible universe is held
and nourished.' There are, he added, 'no atheists
and skeptics in regard to this knowledge.' Nature is
reality, Thoreau said – distinct from the 'Arabian
nights entertainments' that humans concoct for
themselves. 'God himself culminates in the present
moment, and will never be more divine in the lapse
of all the ages. And we are enabled to apprehend at
all what is sublime and noble only by the perpetual
instilling and drenching of the reality that surrounds us.' That drenching could come in the
woods around Walden, but better in true wilderness. On his trip to Mt Katahdin, Thoreau looked
around at the uncut miles and said: 'Perhaps where
our wild pines stand and leaves lie on the forest floor
in Concord, there were once reapers, and husband
men planted grain; but here not even the surface
had been scarred by man . . . It was a specimen of

what God saw fit to make this world.' The earth is a museum of divine intent.

Simply saying that we apprehend God in nature, however, is just a beginning. It may be true, as a mystic once contended, that most people, sometime in their lives, are moved by natural beauty to a 'mood of heightened consciousness' in which 'each blade of grass seems fierce with meaning,' but the question is: what meaning? 'All nature,' contended another mystic a century ago, 'is the language in which God expresses his thought.' Very well, but what thought is that?

The chief lesson is that the world displays a lovely order, an order comforting in its intricacy. And the most appealing part of this harmony, perhaps, is its permanence – the sense that we are part of something with roots stretching back nearly forever, and branches reaching forward just as far. Purely human life provides only a partial fulfillment of this desire for a kind of immortality. As individuals, we can feel desperately alone: we may not have children, or we may not care much for how they have turned out; we may not care to trace ourselves back through our parents; some of us may even be general misanthropes, or feel that our lives are unimportant, brief, and hurried rushes toward a final emptiness. But the earth and all its processes – the sun growing plants, flesh feeding on these plants, flesh decaying to nourish more plants, to name just one cycle – gives us some sense of a more enduring role. The poet Robinson Jeffers, a deeply pessimistic man with regard to the human condition, once wrote, 'The parts change and pass, or die, people and races and rocks and stars; none of them seems to me important in itself, but only the whole . . . It seems to me that this whole alone is worthy of a deeper sort of love; and that there is peace, freedom, I might say a kind of salvation . . .'

John Muir expressed this sense of immortality best. Born to a stern Calvinist father who used a belt to help him memorize the Bible, Muir eventually escaped to the woods, traveling to the Yosemite Valley of California's Sierra Nevada. The journal of his first summer there is filled with a breathless joy at the beauty around him. Again and again in that Sierra June, 'the greatest of all the months of my life,' he uses the word 'immortality,' and he uses it in a specific way, designed to contrast with his father's grim and selfish religion. Time ceases to have its normal meaning in those hills: 'Another glorious Sierra day in which one seems to be dissolved and sent pulsing onward we know not where.

Life seems neither long nor short, and we take no more heed to save time or make haste than do the trees and stars. This is true freedom, a good practical sort of immortality.' In a mood like this, space is no more imposing a boundary than time: 'We are now in the mountains, and they are now in us, making every nerve quiet, filling every pore and cell of us. Our flesh-and-bone tabernacle seems transparent as glass to the beauty around us, as if truly an inseparable part of it, thrilling with the air and trees, streams and rocks, in the waves of the sun – a part of all nature, neither old nor young, sick nor well, but immortal.'

Still, moving as it is, all this remains slightly vague, transcendental. For Burroughs and for Muir and for Thoreau, God didn't have a name or a doctrine. For many of us in the West this fuzzy notion of God is all there is, just as for many others God is all too obvious in his likes and his dislikes. In fact, to the degree that our dominant Judeo-Christian tradition is seen as saying anything about nature, it is usually seen as antienvironmentalist, as elevating man above all others. The Genesis story, with its emphasis upon dominion ('Fill the earth and subdue it, and have dominion over the fish of the sea and over the birds of the air and over every living thing that moves upon the earth'), appears the perfect rationale for cutting down forests, running roads through every wild place, killing off snail darters. The biblical tradition, Joseph Campbell says, is the 'socially-oriented mythology' of a mobile people, as opposed to the nature-oriented mythology of an earth-cultivating society. Therefore, we control nature, or try to. In an influential essay written at the height of the environmental movement, Lynn White, Jr, said that Christianity bears 'an immense burden of guilt' for the ecological crisis; to get some sense of his meaning requires only a trip to Utah, where the state motto is 'Industry' and the Mormons have made a great project of subduing nature, erecting some towns in places so barren and dry and steep that only missionary zeal to conquer the wild could be the motivation.

But Christianity was long the bulwark of slavery, too; indeed, one could make at least as convincing an argument from the text that the Bible countenances chattel bondage as that it urges the rape of the land. Both rely on narrow readings of short passages; when the Bible is read as a whole, I think, the opposite messages resound, though we have been slow to hear them. For every passage like the one in

Genesis there is a verse counseling moderation, love of land. In recent years, many theologians have contended that the Bible demands a careful 'stewardship' of the planet instead of a careless subjugation, that immediately after giving man dominion over the earth God instructed him to 'cultivate and keep it.' But actually, I think, the Scriptures go much deeper. The Old Testament contains in many places, but especially in the book of Job, one of the most far-reaching defenses ever written of wilderness, of nature free from the hand of man. The argument gets at the heart of what the loss of nature will mean to us.

Job is, of course, the story of a just and prosperous man. The devil wagers God that Job's piety is merely a function of his success; bring him down and he will curse you, he says. God agrees to the bet, and soon Job is living on a dunghill on the edge of town, his flesh a mass of oozing sores, his children dead, his flock scattered, his property gone. He refuses to curse God, but he does demand a meeting with him and an explanation of his misfortune. Job refuses to accept the reasoning of his orthodox friends – that he has unknowingly sinned and is therefore being punished. Their view, that all the earth revolves around man, and every consequence is explained by man's action, doesn't satisfy Job: he knows he is innocent.

Finally, God arrives, a voice from the whirlwind. But instead of engaging in deep metaphysical discussion he talks at some length about nature, about concrete creation. 'Where were you when I laid the earth's foundation?' he asks. In an exquisite poem he lists his accomplishments, his pride in his creation always evident. Was Job there when he 'put the sea behind closed doors?' Job was not; therefore Job could not hope to understand many mysteries, including why rain falls 'on land where no one lives, to meet the needs of the lonely wastes and make grass sprout upon the ground.' God seems to be insisting that we are not the center of the universe, that he is quite happy if it rains where there are no people – that God is quite happy with *places* where there are no people, a radical departure from our most ingrained notions.

The end of the book contains descriptions of Behemoth and Leviathan, two creatures God has made and constrained. 'Behold now Behemoth,' booms God. 'He eateth grass as an ox. Lo now, his strength is in his loins. And his force is in the muscles of his belly. He moveth his tail like a cedar . . . His bones are as tubes of brass. His limbs are like bars of iron . . . Behold, if a river overflow he trembleth not. He is confident, though Jordan swell even to his mouth. Shall any take him when he is on the watch, or pierce through his nose with a snare?' The answer, clearly, is no; the message, though not precisely an answer to Job's plaint, is that we may not judge everything from our point of view – that all nature is not ours to subdue.

There are some who have heard that message, even as most of the Western world has gone along its prideful way. Among the company of Christian saints, not one is more beloved than Francis of Assisi. We all have a mental image of him, usually that of a man in a brown robe whose shoulders and arms are covered with birds. His pastoral vision was not entirely unprecedented: for at least the first five centuries of the Church, the dominant Christian symbol had been Christ as the Good Shepherd instead of Christ on the Cross. And, granted, Francis's understanding of the importance of nature was somewhat different from ours – because water was used in baptism, says his biographer William Armstrong, Francis took pains not to tread where he had emptied his washbasin. But his essential idea was not baroque: Just as God had sent Jesus to manifest him in human form, so too he represented himself in birds and flowers, streams and boulders, sun and moon, the sweetness of the air. Holding a small duck in his hand, wrote Bonaventure, Francis was in religious ecstasy: 'He beheld in fair things Him who is the most fair.'

Wild nature, then, has been a way to recognize God and to talk about who he is – even, as in Job, a way for God to talk about who he is. How could it be otherwise? What else is, or was, beyond human reach? In what other sphere could a deity operate freely? It is not chance that every second hymn in the hymn book rings with the imagery of the untouched outdoors. 'All thy works with joy surround thee, Earth and Heaven reflect thy rays,' we sing to Beethoven's 'Ode to Joy.' Sheep and harvests and the other common motifs of the Bible are not just metaphors; they are also the old reality of the earth, a place where people depended for both life and meaning on the nature they found around them. 'We plow the fields and scatter the good seed on the land, But it is fed and watered by God's almighty hand. He sends the snow in winter, The warmth to swell the grain, The breezes and the sunshine, And soft refreshing rain. All good gifts around us Are sent from heaven above.'

*

So what will the end of nature as we have known it mean to our understanding of God and of man? The important thing to remember is that the end of nature is not an impersonal event, like an earthquake. It is something we humans have brought about through a series of conscious and unconscious choices: *we* ended the natural atmosphere, and hence the natural climate, and hence the natural boundaries of the forests, and so on. In so doing, we exhibit a kind of power thought in the past to be divine (much as we do by genetically altering life).

We as a race turn out to be stronger than we suspected – much stronger. In a sense we turn out to be God's equal – or, at least, his rival – able to destroy creation. This idea, of course, has been building for a while. 'We became less and less capable of seeing ourselves as small within creation, partly because we thought we could comprehend it statistically, but also because we were becoming creators, ourselves, of a mechanical creation by which we felt ourselves greatly magnified,' writes the essayist Wendell Berry. 'Why, after all, should one get excited about a mountain when one can see almost as far from the top of a building, much farther from an airplane, farther still from a space capsule?' And our atomic weapons obviously created the *possibility* that we could exercise godlike powers.

But the possibility is different from the fact. We actually seem to have recognized the implications of nuclear weapons, and begun to back away from them – an unprecedented act of restraint. In our wholesale alteration of nature, though, we've shown no such timidity. And just as challenging one's parents and getting away with it rocks one's identity, so must this. Barry Lopez reports that the Yupik Eskimos refer to us Westerners 'with incredulity and apprehension as "the people who change nature."' When changing nature means making a small modification in what we have found – a dam across a river – it presents few philosophical problems. (It presents some, especially when the river is a beautiful one, but they tend not to be ultimate problems.) When changing nature means changing everything, then we have a crisis. We are in charge now, like it or not. As a species we are as gods – our reach global.

And God has not stopped us. The possibilities – if there is or was any such thing as God, the eternal, the divine – include at least the following. God thoroughly approves of what we have done; it is our destiny. God doesn't approve, but is powerless to do anything about it, either because he is weak or because he has created us with free will. Or God is uninterested, or absent, or dead.

That last option is not a new formulation, of course. Nietzsche said some time ago that God was dead, and a lot of people began to agree with him after the Holocaust. The Holocaust and what I am calling the end of nature are not comparable events: the latter is an idea, like the closing of the frontier, and, at least for the moment, has less physical reality. But it may have similar faith-shattering effects. To many whose faith was built on God's covenant with the Israelites, on his promise to protect them, the Holocaust crushed belief or altered it enormously. For some Jewish thinkers, wrote the theologian Marc Ellis, 'the Holocaust represents the severing of the relationship between God and person, God and community, God and culture. The lesson of the Holocaust is that humanity is alone and there is no meaning in life outside of human solidarity.' (And human solidarity, of course, is eternally thrown into question by the Holocaust.) In a similar fashion, for those of us who have tended to locate God in nature – who, say, look upon spring as a sign of his existence and a clue to his meaning – what does it mean that we have destroyed the old spring and replaced it with a new one of our own devising? Why did he not stop us? Why did he allow it?

Perhaps it is all for the best, a break with some Druidic past. But it seems infinitely sad. And it seems to feed on itself, unlike the Holocaust, whose lessons maybe just possibly did increase the chances of human caring. How are we to be humble in any way if we have taken over as creators? Thoreau once stood in the woods watching 'an insect crawling amid the pine needles on the forest floor, and endeavoring to conceal itself from my sight.' It reminded him, he said (and Thoreau was not an especially humble man) of 'the greater Benefactor and Intelligence that stands over me, the human insect.' But what stands over us?

Religion will not end – far from it. We are probably in for a siege of apocalyptic and fanatic creeds. But a certain way of thinking about God – a certain language by which to describe the indescribable – will disappear. The stern God of Muir's father talked constantly of sin and condemnation, and in booming, angry tones. Muir's God spoke to him in the rush of water across the rocks and the cry of the jays around his camp. They were different Gods. 'If we have a wonderful sense of the divine, it is because we live amid such awesome magnificence,' wrote religious scholar Thomas Berry. 'If we lived on the moon, our mind and emotions, our speech, our imagination, our sense of the divine would all reflect the desolation of the lunar landscape.'

And even if we manage to control the physical effects of our actions – if we come to live in a planet-size park of magnificent scenery – our sense of the divine will change. It will be, at best, the difference between a zoo and a wilderness. The Bronx Zoo has done a wonderful job of exchanging cages for wide, grassy fields, but even though the antelope have room to get up to speed and the zebra wander as a striped herd, it never crosses your mind that you are actually in the bush instead of the Bronx. We live, all of a sudden, in an Astroturf world, and though an Astroturf world may have a God, he can't speak through the grass, or even be silent through it and let us hear.

'Science,' of course, replaced 'God' as a guiding concept for many people after Darwin. Or, really, the two were rolled up into a sticky ball. To some degree this was mindless worship of a miracle future, the pursuit of which has landed us in the fix we now inhabit. I was browsing the other day through a volume from the 1950s edited by the eminent astronomer Harlow Shapley. Called *A Treasury of Science*, it is filled with the wisdom of the ages, essays dating back to Hippocrates. But it also includes one example of the wisdom of our particular age, a thirteen-page treatise in which one Roger Adams forecasts the wonderful epoch ahead in *Man's Synthetic Future*. Chemists, he predicts, will replace natural products with 'new, better, and cheaper compounds' of their own creation. 'An official of the wool industry made a statement recently that the demand for wool as a fabric will never be replaced,' Adams scoffs. 'These words were spoken by one completely unfamiliar with the potentialities of chemical research.' Leather, too: 'With durable, moisture-absorbing plastics, the problem of synthetic shoe uppers will be solved.' On and on he goes, through the wonders of DDT, the high hopes for chemicals that will 'effectively kill the crabgrass in the bluegrass lawn,' and a hundred other miracles. 'Today life is mechanized, electrified, abundant, easy, because of the push-button era,' he concludes. 'In the future citizens will more effectively farm the land and the seas; obtain necessary minerals from the oceans; clothe themselves from the coal and oil . . . be cured of any ailments by a variety of drugs and medicinals; be happy, healthy, and kittenish at a hundred years of age; and perhaps attend interplanetary football matches in the Rose Bowl.'

Not everyone who fell in love with science was such a glib Dacron worshiper. An example, typical of a certain strain, was Donald Culross Peattie, a nature writer prominent in the years around World War II. (Though his work has been largely forgotten, one of his books, *An Almanac of Moderns*, was chosen by a book club as the American volume written in the three years preceding 1940 that was 'most likely to become a classic.') Peattie defends the scientific faith as fiercely as any man could: 'What is the force, the discipline, the brotherhood bound by vows to the pursuit of incorruptible truth, which proves every step, is forever returning to verify, will abandon any cherished tenet the moment it is not convincing?' he asks. 'What is it that works all the modern miracles, has put the practicality into compassion for the suffering, has unchained men from their superstitions, has endured persecution and martyrdom, and still knows no fear?' Well, science, of course. But science is only a method of getting at truth; it's the truth that matters. And in Peattie's case, and in many others, the truth that emerged was – nature.

Peattie lived at the moment when ecological understanding was beginning to break through, and he found great comfort and safety in the repeating patterns of nature, in the constant elements of the periodic table that make up the earth and the stars. 'If by "supreme command" I may express an order in nature that a man can understand and revere, then that command, that order, has always been there. In fact, it is nature itself, revealed in science.' Biologists, astronomers, and physicists, 'those who have looked most deeply,' were the 'surest, serenest' men that Peattie knew, because they understood that 'the immutable order of Nature is on our side. It is on the side of life.'

The hope that science could replace religion as a way for human beings to cope with the world, then, was really a hope that 'nature' could replace 'God' as a source of inspiration and understanding. Harmony, permanence, order, and an idea of our place in that order – scientists searched for all that as diligently as Job, with their unceasing attention to the 'web of life' and the grand cycle of decay and rebirth. But nature, it turned out, was fragile:

men could turn it on its head so that it was no longer 'immutable' and no longer 'on the side of life.' The atom bomb proved that, by combining some of the elements in a new and interesting way that clearly held the possibility of wiping out most life. The useful ecological insight that, in Peattie's words, 'it is even good to die, since death is a natural part of life' clearly didn't apply to atomic annihilation, nor, I think, does it apply to death in a world where the natural cycles have been so altered. What is a 'natural part of life' in an unnatural world? How, if the seasons are no longer inevitable, can we accept the inevitability, and even the beauty, of death?

Scientists may argue that natural processes still rule – that the chemical reactions even now eating away the ozone or absorbing the earth's reflected heat are proof that nature is still in charge, still our master. And some physicists have always talked about God in the interstices of the atom, or in the mysteries of quantum theory, or, more recently, says Robert Wright, in his *Three Scientists and Their Gods*, in stitches of DNA and other bits of 'information.' To all but the few hundred people who really understand the math, though, this is a minor and secondhand comfort, an occult, esoteric knowledge. We draw our lessons from what we can see and feel and hear around us. The nature that matters is not the whirling fuzziness of electrons and quarks and neutrinos, which will continue unchanged; it is not the vast and strange worlds and fields and fluxes that scientists can find with their telescopes. The nature that matters is the temperature, and the rain, and the leaves turning color on the maples, and the raccoons around the garbage can.

We can no longer imagine that we are part of something larger than ourselves – that is what all this boils down to. We used to be. When we were only a few hundred million, or only a billion or two, and the atmosphere had the composition it would have had with or without us, then even Darwin's revelations could in the end only strengthen our sense of belonging to creation, and our wonder at the magnificence and abundance of that creation. And there was the possibility that something larger than us – Francis's God, Thoreau's Benefactor and Intelligence, Peattie's Supreme Command – reigned over us. We were as bears – we slept less, made better tools, took longer to rear our young, but we lived in a world that we found made for us, by God, or by physics and chemistry and biology, just as bears live in a world they find waiting for them. But now *we* make that world, affect its every operation (except a few – the alteration of day and night, the spin and wobble and path of the planet, the most elementary geologic and tectonic processes).

As a result, there is no one by our side. Bears are now a distinctly different order of being, creatures in our zoo, and they have to hope we can figure out a way for them to survive on our hot new planet. By domesticating the earth, even though we've done it badly, we've domesticated all that live on it. Bears hold more or less the same place now as golden retrievers. And there is nobody above us. God, who may or may not be acting in many other ways, is not controlling the earth. When he asks, as he does in Job, 'Who shut in the sea with doors . . . and prescribed bounds for it?' and 'Who can tilt the waterskins of the heavens?' we can now answer that it is us. Our actions will determine the level of the sea, and change the course and destination of every drop of precipitation. This is, I suppose, the victory we have been pointing to at least since the eviction from Eden – the domination some have always dreamed of. But it is the story of King Midas writ large – the power looks nothing like what we thought it would. It is a brutish, cloddish power, not a creative one. We sit astride the world like some military dictator, some smelly Papa Doc – we are able to wreak violence with great efficiency and to destroy all that is good and worthwhile, but not to exercise power to any real end. And, ultimately, that violence threatens us. Forget the interplanetary Rose Bowl; 'man's synthetic future' has more to do with not going out in the sun for fear of cancer.

But the cancer and the rising sea level and the other physical effects are still in the future. For now, let's concentrate on what it feels like to live on a planet where nature is no longer nature. What is the sadness about?

In the first place, merely the knowledge that we screwed up. It may have been an inevitable divorce: man, so powerful, may not have been meant to live forever within the constraints of nature. It may have been an inevitable progression – man growing up to be stronger than his mother, nature. But even inevitable passages such as these are attended by grief. Ambition, growth, take us away from old comforts and assurances. We are used to the idea that something larger than we are and not of our own making surrounds us, that there is a world of man and a world of nature. And we cling to that idea in part because it makes that world of men easier to deal with. E. B. White, in one of his last essays, written from his saltwater farm near Mt Desert in Maine, said that 'with so much disturbing our lives and clouding our future . . . it is hard to foretell what is going to happen.' But, he continued, 'I know one thing that *has* happened: the willow by the brook has slipped into her yellow dress, lending, along with the faded pink of the snow fence, a spot of color to the vast gray-and-white world. I know, too, that on some not too distant night, somewhere in pond or ditch or low place, a frog will awake, raise his voice in praise, and be joined by others. I will feel a whole lot better when I hear the frogs.' There may still be frogs – there may be *more* frogs, for all I know – but they will be messengers not from another world, whose permanence and routine can comfort us, but from a world that is of our own making, as surely as Manhattan is of our own making. And while Manhattan has many virtues, I have never heard anyone say that its sounds make you feel certain that the world, and you in it, are safe.

Anyway, I don't think that this separation was an inevitable divorce, the genetically programmed growth of a child. I think it was a mistake, and that consciously or unconsciously many of us realize it was a mistake, and that this adds to the sadness. Many have fought to keep this day from coming to pass – fought local battles, it is true, perhaps without realizing exactly what was at stake, but still understanding that the independent world of nature was gravely threatened. By the late 1960s an 'environmental consciousness' had emerged, and in the 1970s and 1980s real progress was being made: air pollution in many cities had been reduced, and wilderness set aside, and Erie, the dead lake, that symbol of ultimate degradation, rescued from the grave.

So there is the sadness of losing something we've begun to fight for, and the added sadness, or shame, of realizing how much more we could have done – a sadness that shades into self-loathing. We, all of us in the First World, have participated in something of a binge, a half century of unbelievable prosperity and ease. We may have had some intuition that it *was* a binge and the earth couldn't support it, but aside from the easy things (biodegradable detergent, slightly smaller cars) we didn't do much. We didn't turn our lives around to prevent it. Our sadness is almost an aesthetic response – appropriate because we have marred a great, mad, profligate work of art, taken a hammer to the most perfectly proportioned of sculptures.

There is also another emotional response – one that corresponds to the cry 'What will I do without him?' when someone vital dies.

I took a day's hike last fall, walking Mill Creek from the spot where it runs by my door to the place where it crosses the main county road near Wevertown. It's a distance of maybe nine miles as the car flies, but rivers are far less efficient, and endlessly follow pointless, time-wasting, uneconomical meanders and curves. Mill Creek cuts some fancy figures, and so I was able to feel a bit exploratory – a budget Bob Marshall. In a strict sense, it wasn't much of an adventure. I stopped at the store for a liverwurst sandwich at lunchtime, the path was generally downhill, the temperature stuck at an equable 55 degrees, and since it was the week before the hunting season opened I didn't have to sing as I walked to keep from getting shot. On the other hand, I had made an arbitrary plan – to follow the creek – and, as a consequence, I spent hours stumbling through overgrown marsh, batting at ten-foot saplings and vines, emerging only every now and then, scratched and weary, into the steeper wooded sections. When Thoreau was on Katahdin, nature said to him, 'I have never made this soil for thy feet, this air for thy breathing, these rocks for thy neighbors. I cannot pity nor fondle thee there, but forever relentlessly drive thee hence to where I *am* kind. Why seek me where I have not called thee, and then complain because you find me but a step-mother?' Nature said this to me on Mill Creek, or at least it said, 'Go home and tell your wife you walked to Wevertown.' I felt I should have carried a machete, or employed a macheteist. (The worst thing about battling through brake and bramble of this sort is that it's so anonymous – gray sticks, green stalks with reddish thorns, none of them to be found in any of the many guides and almanacs on my shelf.) And though I started the day with eight dry socks, none saw noon in that pleasant state.

If it was all a little damp and in a minor key, the sky was nonetheless bright blue, and rabbits kept popping out from my path, and pheasants fired up between my legs, and at each turning some new gift appeared: a vein of quartz, or a ridge where the maples still held their leaves, or a pine more than three feet in diameter that beavers had gnawed all the way around and halfway through and then left standing – a forty-foot sculpture. It was October, so there weren't even any bugs. And always the plash of the stream in my ear. It isn't Yosemite, the Mill Creek Valley, but its small beauties are absorbing, and one can say with Muir on his mountaintop, 'Up here all the world's prizes seem as nothing.'

And so what if it isn't nature primeval? One of our neighbors has left several kitchen chairs along his stretch of the bank, spaced at fifty-yard intervals for comfort in fishing. At one old homestead, a stone chimney stands at either end of a foundation now filled by a graceful birch. Near the one real waterfall, a lot of rusty pipe and collapsed concrete testifies to the old mill that once stood there. But these aren't disturbing sights – they're almost comforting, reminders of the way that nature has endured and outlived and with dignity reclaimed so many schemes and disruptions of man. (A mile or so off the creek, there's a mine where a hundred and fifty years ago a visionary tried to extract pigment for paint and pack it out on mule and sledge. He rebuilt after a fire; finally an avalanche convinced him. The path in is faint now, but his chimney, too, still stands, a small Angkor Wat of free enterprise.) Large sections of the area were once farmed; but the growing season is not much more than a hundred days, and the limits established by that higher authority were stronger than the (powerful) attempts of individual men to circumvent them, and so the farms returned to forest, with only a dump of ancient bottles or a section of stone wall as a memorial. (Last fall, though, my wife and I found, in one abandoned meadow, a hop vine planted at least a century before. It was still flowering, and with its blossoms we brewed beer.) These ruins are humbling sights, reminders of the negotiations with nature that have established the world as we know it.

Changing socks (soaking for merely clammy) in front of the waterfall, I thought back to a recent spring, when a record snowfall melted in only a dozen or so warm April days. A little to the south, an inflamed stream washed out a highway bridge, closing the New York Thruway for months. Mill Creek filled till it was a river, and this waterfall, normally one of those diaphanous-veil affairs, turned into a cataract. It filled me with awe to stand there then, on the shaking ground and think, This is what nature is capable of.

But as I sat there this time, and thought about the dry summer we'd just come through, there was nothing awe-inspiring or instructive, or even lulling, in the fall of the water. It suddenly seemed less like a waterfall than like a spillway to accommodate the overflow of a reservoir. That didn't decrease its beauty, but it changed its meaning. It has begun or will soon begin to rain and snow when the particular mix of chemicals we've injected into the atmosphere adds up to rain or snow – when they make it hot enough over some tropical sea to form a cloud and send it this way. I had no more control, in one sense, over this process than I ever did. But it felt different, and lonelier. Instead of a world where rain had an independent and mysterious existence, the rain had become a subset of human activity: a phenomenon like smog or commerce or the noise from the skidder towing logs on Cleveland Road – all things over which I had no control, either. The rain bore a brand; it was a steer, not a deer. And that was where the loneliness came from. There's nothing there except us. There's no such thing as nature anymore – that other world that isn't business and art and breakfast is now not another world, and there is nothing except us alone.

At the same time that I felt lonely, though, I also felt crowded, without privacy. We go to the woods in part to escape. But now there is nothing except us and so there is no escaping other people. As I walked in the autumn woods I saw a lot of sick trees. With the conifers, I suspected acid rain. (At least I have the luxury of only suspecting; in too many places, they *know*). And so who walked with me in the woods? Well, there were the presidents of the Midwest utilities who kept explaining why they had to burn coal to make electricity (cheaper, fiduciary responsibility, no *proof* it kills trees) and then there were the congressmen who couldn't bring themselves to do anything about it (personally favor but politics the art of compromise, very busy with the war on drugs) and before long the whole human race had arrived to explain its aspirations. We like to drive, they said, air conditioning is a necessity nowadays, let's go to the mall. By this point, the woods were pretty densely populated. As I attempted to escape, I slipped on another rock, and in I went again. Of course, the person I was fleeing most fearfully was myself, for I drive (I drove forty thousand miles one year), and I'm burning a collapsed barn behind the house next week because it is much the cheapest way to deal with it, and I live on about four hundred times what Thoreau conclusively proved was enough, so I've done my share to take this independent, eternal world and turn it into a science-fair project (and not even a good science-fair project but a cloddish one, like pumping poison into an ant farm and 'observing the effects').

The walk along Mill Creek, or any stream, or up any hill, or through any woods, is changed forever – changed as profoundly as when it shifted from pristine and untracked wilderness to mapped and deeded and cultivated land. Our local shopping mall now has a club of people who go 'mall walking' every day. They circle the shopping center en masse – Caldor to Sears to J. C. Penney, circuit after circuit with an occasional break to shop. This seems less absurd to me now than it did at first. I like to walk in the outdoors not solely because the air is cleaner but because outdoors we venture into a sphere larger than ourselves. Mall walking involves too many other people, and too many purely human sights, ever to be more than good-natured exercise. But now, out in the wild, the sunshine on one's shoulders is a reminder that man has cracked the ozone, that, thanks to us, the atmosphere absorbs where once it released.

The greenhouse effect is a more apt name than those who coined it imagined. The carbon dioxide and trace gases act like the panes of glass on a greenhouse – the analogy is accurate. But it's more than that. We have built a greenhouse, *a human creation*, where once there bloomed a sweet and wild garden.